Table of Conte

Chapter 1 Bird Nests and More.....................4

Chapter 2 From Holes in the Ground to High Cliffs..6

Chapter 3 Nests of Sand and Bubbles..............14

Chapter 4 Nests for Mammals......................18

Glossary..22

To Find Out More..................................23

Index...24

Words that appear in the glossary are printed in **boldface** type the first time they occur in the text.

Chapter 1

Bird Nests and More

What do you think of when you hear the word *nests*? Are nests only for birds?

Many other animals make nests, too. Some fish build nests in water. Chimpanzees make nests in trees.

WHERE ANIMALS LIVE

Why Animals Live in Nests

By Valerie J. Weber

Reading consultant:
Susan Nations, M.Ed., *author/literacy coach/consultant in literacy development*

Science and curriculum consultant:
Debra Voege, M.A., *science curriculum resource teacher*

WEEKLY READER®
PUBLISHING

Please visit our web site at www.garethstevens.com.
For a free color catalog describing our list of high-quality books, call 1-800-542-2595 (USA) or 1-800-387-3178 (Canada). Our fax: 1-877-542-2596

Library of Congress Cataloging-in-Publication Data
Weber, Valerie.
    Why animals live in nests / by Valerie J. Weber.
       p. cm. — (Where animals live)
    Includes bibliographical references and index.
    ISBN-10: 0-8368-8796-4   ISBN-13: 978-0-8368-8796-9 (lib. bdg. : alk. paper)
    ISBN-10: 0-8368-8803-0   ISBN-13: 978-0-8368-8803-4 (softcover)
    1. Birds—Nests—Juvenile literature. 2. Animals—Habitations—Juvenile literature. 3. Nests—Juvenile literature. I. Title.
    QL675.W37   2008
    591.56'4—dc22   2007041627

This edition first published in 2008 by
Weekly Reader® Books
An Imprint of Gareth Stevens Publishing
1 Reader's Digest Road
Pleasantville, NY 10570-7000 USA

Copyright © 2008 by Gareth Stevens, Inc.

Senior Managing Editor: Lisa M. Guidone
Senior Editor: Barbara Bakowski
Creative Director: Lisa Donovan
Senior Designer: Keith Plechaty
Production Designer: Amy Ray, *Studio Montage*
Photo Researcher: Diane Laska-Swanke

Photo Credits: Cover © John Eastcott and Yva Momatiuk/Getty Images; pp. 1, 3, 4 © Photodisc; p. 5 © Solvin Zankl/naturepl.com; p. 7 © Lenice Harms/Shutterstock; p. 8 © Fletcher & Baylis/Photo Researchers, Inc.; p. 9 © Tom and Pat Leeson; p. 10 © Hamiza Bakirci/Shutterstock; p. 11 © Tom Mangelsen/naturepl.com; p. 12 © Danita Delimont/Alamy; p. 13 © Ron Niebrugge/Alamy; p. 15 © Dwight Kuhn; p. 16 © Doug Perrine/naturepl.com; p. 17 © Lynn M. Stone/naturepl.com; p. 19 © coko/Shutterstock; p. 20 © Bruce Coleman Inc./Alamy; p. 21 © Ingo Arndt/Minden Pictures

All rights reserved. No part of this book may be reproduced, stored in a retrieval system, or transmitted in any form or by any means, electronic, mechanical, photocopying, recording, or otherwise, without the prior written permission of the copyright holder.

Printed in the United States of America

1 2 3 4 5 6 7 8 9 10 09 08 07

A skua (SKYOO-uh) flies low over a nest of gentoo (JEN-too) penguins.

Different animals make different kinds of nests. All nests have the same purposes, though. Animals use their nests for shelter. Some animals raise their babies in nests. Animals make their nests in places that are safe from **predators**. Predators are animals that eat other animals.

Chapter 2

From Holes in the Ground to High Cliffs

Most birds build nests. Some nests are as simple as a small hole in the ground. That kind of nest is called a scrape nest. Other nests are huge **structures** built high in trees. Some birds use rocks, mud, or sticks to build nests. Others use feathers, twigs, spiderwebs, grass, or spit to make their homes.

People build different kinds of homes in different places. In hot places, people build homes that keep them cool. In cold places, they build houses that keep them warm. Their homes suit their **environment**.

Shore breezes cool this house built on stilts over water.

The type of nest a bird builds depends on where the animal lives. Birds that live on the seashore often build their nests in sand or stones. The birds move the stones or sand to make a small hollow. The hollow looks like a little bowl.

The arctic tern makes its nest in a rocky area. The bird's spotted eggs look like stones, making them hard for predators to see.

You may have seen robins near your house. Robins usually build nests in thick bushes or trees. The leaves help hide their nests. Predators cannot find the nests easily.

This robin's nest is well hidden in a flowering tree.

Robins pull up thick grasses and find twigs to make their nests. The birds use mud to hold the grass and twigs together. To make the nests soft for their eggs, they line the nests with fresh grass.

Robins "glue" their nests together with mud.

Eagles need a wide area to hunt in. From high in the air, they can see small animals on the ground. Eagles catch the animals and feed them to their **chicks**, or baby birds.

An eagle soars high above the ground to look for food. The eagle has very sharp eyesight.

Eagles build their nests in tall trees or at the edge of cliffs. No other animal eats eagles, so they can nest out in the open.

Both male and female eagles build the nest. First, the adult eagles gather large sticks.

A bald eagle nests on a high cliff.

Then the eagles weave the sticks together. Finally, the adults line the nest with moss, grass, and needles from pine trees. Some eagles' nests are very big and strong. A grown-up person could stand in a bald eagle's nest without breaking it!

Eagles build huge nests in the tops of tall trees.

Chapter 3

Nests of Sand and Bubbles

Some fish and **reptiles** also build nests for their eggs and young. Reptiles are animals with dry, scaly skin, such as lizards and turtles.

Some fish simply scrape away small pebbles from the seafloor and stick their eggs on bare rocks. Other fish build nests from small plants.

Some fish also use bubbles to form their nests. They blow bubbles of air from their mouths. Then they use their spit to stick the bubbles together. They put a fish egg in each bubble. The male fish guard the bubbles from predators until the eggs **hatch**.

Some fish build floating bubble nests to hold their eggs.

Many reptiles also build nests for their eggs. Some do not stay around and wait for the eggs to hatch, however.

One of those reptiles is the loggerhead turtle. This big turtle spends most of its life in the sea.

When a loggerhead turtle is ready to lay eggs, it swims back to the beach where it was born.

When it is time for the female turtle to give birth, she drags herself onto a beach. Using her big **flippers**, she digs a hole in the warm sand. There, she lays about one hundred eggs and covers them with sand. Then she goes back to the ocean. The mother turtle never returns to the eggs—or the baby turtles.

The female loggerhead crawls onto the beach. She digs a nest for her eggs in the sand.

Chapter 4

Nests for Mammals

Mammals are very different from loggerhead turtles. Mammals are warm-blooded animals that make milk for their babies. Some mammals have babies in their nests and stay to take care of them.

Cottontail rabbits are mammals that live in North America. You might have seen one near your home.

They have large eyes, long ears, and strong back legs. These rabbits get their name from their tail—a small puff of white fur.

Female rabbits are called **does** (doze). They build nests in places that are hidden from predators. You can sometimes find a rabbit's nest under a bush or a log. Sometimes rabbits make their nests in tall grass.

A cottontail rabbit scrapes a hole in the ground beneath a bush or a log in tall grass.

A doe gathers grass to build a nest. She also takes fur from her body and lines the nest with it. Baby rabbits have no fur when they are born. The soft fur in the nest protects their skin.

A cottontail rabbit usually has four or five babies at a time.

Chimpanzees are also mammals. The mother chimp builds a new nest every night. She breaks off tree branches. Then she piles them between the trunk of the tree and two tree **limbs**. There, mother chimp and her baby sleep.

Around the world, human babies also sleep near their mothers in their homes every night.

Each night, a mother chimp makes a sleep nest for her and her baby.

Glossary

chicks: baby birds

does: adult females of some animals

environment: the area where a person, an animal, or a plant lives

flippers: broad, flat parts of an animal that are used for swimming

hatch: to come out of an egg

limbs: large branches of trees

mammals: animals that are warm-blooded and that have a backbone. They feed their babies milk made in their bodies.

predators: animals that hunt and eat other animals

reptiles: animals that have skin covered with scales or bony plates. Reptiles lay eggs.

structures: things that are built or made

To Find Out More

Books
Birds and Their Nests. Linda Tagliaferro (Pebble Plus)

Bird, Nests, and Eggs. Mel Boring (T&N Children's Publishing)

How and Why Birds Build Nests. How and Why (series). Elaine Pascoe (Creative Teaching Press)

Squirrels and Their Nests. Animal Homes (series). Martha E. Rustad (Coughlan Publishing)

Web Sites
The Birdhouse Network
www.birds.cornell.edu/birdhouse/nestboxcam
Watch videos of birds building and living in their nests.

PBS: Nature
www.pbs.org/wnet/nature/turtles/navigate.html
Learn more about how turtles find the beaches where they lay their eggs.

Ranger Rick
www.nwf.org/kidzone/kzPage.cfm?siteId=3&departmentId=82&articleId=889
Learn how alligators build their nests and how their eggs stay warm.

Publisher's note to educators and parents: Our editors have carefully reviewed these web sites to ensure that they are suitable for children. Many web sites change frequently, however, and we cannot guarantee that a site's future contents will continue to meet our high standards of quality and educational value. Be advised that children should be closely supervised whenever they access the Internet.

Index

arctic tern 8
bird 4, 5, 6, 8–13
bubble nest 15
chimpanzee 4, 21
doe 19–20
eagle 11–13
eggs 8, 10, 16–17
environment 7
fish 4, 14–15
flippers 17
hatching 15, 16

mammal 18–21
penguins 5
predator 5, 9, 15, 19
rabbit, cottontail 18–20
reptile 14, 16–17
robin 9–10
sand 8, 17
seashore 8, 17
skua 5
turtle, loggerhead 16–17

About the Author

A writer and editor for more than twenty-five years, Valerie Weber especially loves working in children's publishing. Her book topics have been endlessly engaging—from the weird wonders of the sea, to the lives of girls during World War II, to the making of movies. She is grateful to her family, including her husband and daughters, and her friends for offering their support and for listening to the odd facts she has discovered during her work. Did you know, for example, that frogs use their eyeballs to push food down into their stomachs?